YOUR KNOWLEDGE HAS VALUE

- We will publish your bachelor's and master's thesis, essays and papers

- Your own eBook and book - sold worldwide in all relevant shops

- Earn money with each sale

Upload your text at www.GRIN.com
and publish for free

Sustainable Farming of Giant Freshwater Prawn (Macrobrachium Rosenbergii) in India

Madhav Karthikeyan

Bibliographic information published by the German National Library:

The German National Library lists this publication in the National Bibliography; detailed bibliographic data are available on the Internet at http://dnb.dnb.de.

ISBN: 9783346991928
This book is also available as an ebook.

Review on sustainable farming of giant freshwater prawn (_Macrobrachium rosenbergii_) in India

Abstract:

Freshwater prawns, commonly known as **Macrobranchium rosenbergii**, are one of the most significant species for aquaculture worldwide. They are widely cultured for their high-quality meat, which is rich in protein, and they are also in high demand in the food industry. Unlike marine shrimps, freshwater prawns have a more extended culture period and face several challenges, making their culture a complex process.

The primary concern in culturing giant freshwater prawns is growth suppression caused by morphological variations within a population. The life cycle of the giant freshwater prawn has three distinct morphological stages: small-clawed, orange-clawed, and blue-clawed individuals. Growth reduction is especially prevalent in male populations due to their aggressive behavior. The social organization of prawns is a crucial factor to consider. In a prawn population, individuals are suppressed in the following order: **small-clawed < orange-clawed < blue-clawed.**

Although culling can be somewhat effective, frequent partial harvesting and segregation of weak and small-clawed individuals can increase profits. This technique ensures that the larger and stronger individuals are not suppressed by their weaker counterparts, leading to better growth rates and higher yields.

This article proposes several techniques for enhancing the yield of scampi farming. These techniques include adjusting culture conditions, feeding practices, and water quality parameters. All of these factors play a significant role in the growth and development of freshwater prawns, and optimizing them can lead to increased yield and profitability.

In conclusion, freshwater prawns are a valuable aquaculture species, and their culture requires careful management to ensure optimal growth and yield. By applying the techniques proposed in this article, farmers can improve their production and meet the growing demand for high-quality freshwater prawns.

Keywords:

Freshwater prawn culture, Advancements, recent trends, Boosting economic yield

1. Introduction:

Freshwater prawn farming has a rich history of captive rearing. However, during the late 1960s, Shao-wen Ling, an FAO official working associate in Malaysia, made a significant breakthrough in freshwater prawn cultivation. Ling discovered that the larval stages of *M.rosenbergii* required brackish water for development to post-larvae, which revolutionized freshwater prawn farming. In 1972, Takuji Fujimura and his Hawaiian team developed hatchery production techniques for commercially producing prawn post-larvae (PL).

According to the FAO, in 2020, scampi farming contributed approximately 234.4 thousand tonnes, 2.5% of the total major crustacean species cultivated worldwide. On the other hand, white-leg shrimp (*Penaeus vannamei*) contributed 52.9% of total crustacean production. The FAO estimates that world aquaculture production will increase by 32.2% by 2030, and scampi farming significantly boosts this growth.

Scampi farming has superior cultivable characteristics such as fast growth, high market demand, hardiness, euryhaline nature, and compatibility with cultivable finfish such as Indian major carps, tilapia, and catfishes, making it incredibly valuable. From the mid-1990s until 2005, *M.rosenbergii* production in India increased dramatically from 178 to 42,870 tonnes. Monoculture of *M.rosenbergii* has achieved yields of 1.0–1.5 tonnes/ha in a cycle lasting 7–8 months.

Freshwater prawn farming has seen impressive growth in India, producing over 30,000 tonnes of water from 35,000 ha in 2002-2003. 38,819 hectares of water were devoted to prawn farming in Andhra Pradesh, followed by 4,744 hectares in West Bengal (Source: MPEDA). Scampi farming has increased significantly in India in recent years, with companies like Coastal Corporation Limited and Baby Marine Ventures driving the industry's transformation. The Indian government has also given significant support to the sector, with initiatives like the National Fisheries Development Board (NFDB) and the National Bank for Agriculture and Rural Development (NABARD) providing financial and technical assistance to farmers.

Overall, freshwater prawn farming and scampi farming, in particular, continue to offer significant opportunities for growth in aquaculture production worldwide, and they are expected to play a vital role in meeting the growing demand for seafood in the coming years. Figure 1 shows the growth in global production of scampi from the year 2000 to 2020

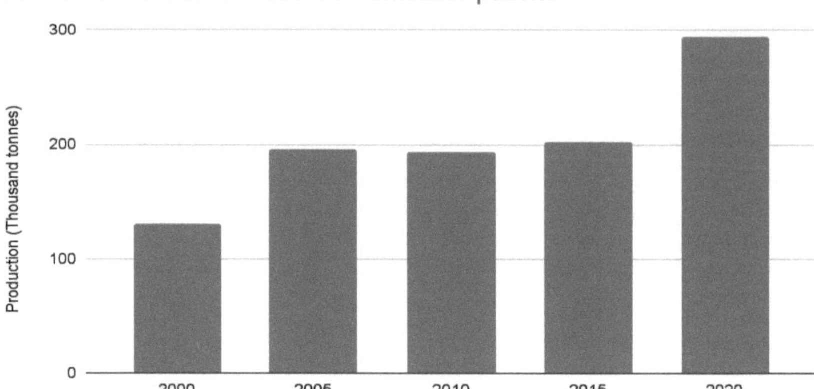

Figure 1 Production trend for Macrobrachium rosenbergii from 2000 to 2020.

2. Barriers in scampi farming:

Scampi farming has become popular recently, but it has some limitations and disadvantages. One of the major problems is that seed availability is restricted to a specific time, and fewer hatcheries are available compared to *Penaeus vannamei* (white-leg shrimp). This makes it difficult for farmers to obtain the necessary seeds to start their farms. Poor post-larvae selection and overstocking are also concerns that can affect the quality of the prawns produced.

Overutilization of groundwater in Nellore has caused stunted growth, according to New et al. This highlights the importance of sustainable water usage in scampi farming. White Tail Disease (WSD) is another issue that causes 100% mortality in larval and post-larval stages of prawns. The causative agent is *Macrobrachium rosenbergii* nodavirus (Mr. NV), and the transmission is vertical. Bacterial disease caused by Vibrio spp and protozoan disease, such as microbial fouling, are also reported in some cases. These diseases can lead to significant losses for farmers and negatively impact the industry.

Handling and packaging are also significant problems in scampi farming. Prawns require specific knowledge to manage appropriately. Mishandling of shrimp can cause increased stress and disease development. Sex segregation of *M.rosenbergeii* using manual labor in Nellore resulted in the development of stress among the individuals, according to Saurabh S and Sahoo PK. This highlights the need for proper training for workers in the industry.

Packaging is also a significant challenge for scampi farmers. Scampi is made of soft flesh, which can deteriorate quickly. Immediate icing of *M.rosenbergii* is necessary for on-site chilling in Nellore, according to Kutty et al. Rapid chilling of prawns is crucial for exporting them in a customer-desirable condition, as their texture becomes soft due to biochemical changes within them. Proper packaging can help preserve the prawns' quality and reduce spoilage during transportation.

These are the major problems faced by scampi farmers worldwide. Addressing these issues is crucial for the sustainable growth of the industry and the production of high-quality prawns.

3. Sustainable scampi farming:

Compared to *Penaeus vannamei* culture, scampi farming is not widely practiced and is only scatteredly practiced in some regions of India and other parts of the world. Inland (freshwater) fisheries contributed almost 51.3% in 2018, while *M.rosenbergii* contributed only 2.5%. Despite this, scampi farming can be sustainable if proper farming procedures are followed to increase production. To achieve sustainable production, it is crucial to have a thorough understanding of the nutritional requirements, growth phase, and water quality parameters of scampi.

Sustainable scampi farming involves implementing practices that minimize environmental impact while producing high-quality, healthy scampi. This means that farmers must take steps to ensure that their farming practices are not causing harm to the surrounding ecosystem. Proper management of water quality parameters, such as pH levels, temperature, and oxygen levels, is essential to maintaining a healthy environment for the scampi to grow.

Furthermore, farmers must also ensure that the scampi receives the appropriate nutrients at the right time to promote optimal growth and development. This involves monitoring the scampi's nutritional requirements and providing them with a balanced diet that meets their needs. Following these procedures can achieve sustainable scampi farming, leading to increased production and a healthier ecosystem.

3.1 Nutritional requirement:

The culture of *Macrobranchium rosenbergii*, commonly known as freshwater prawns, thrives on adequate nutrition and feeding practices. To ensure their growth and maintenance, providing them with a balanced diet that includes animal- and plant-based protein sources is crucial. The variable nutritional requirements essential for their growth are listed in Table 1.1.

Protein is a vital component of the freshwater prawn's diet, and their nutritional requirement for protein is between 35-40%. An energy level of 3.2 kcal per gram of feed is essential to meet their metabolic needs. The protein-energy ratio of 125-130 mg of protein per Kcal of energy is critical for their growth and development.

Dietary fibers are an essential component of the freshwater prawn's diet, and they require about 30% of dietary fibers in their feed. Vitamins, such as vitamin C, are also necessary for their growth. They require 60-150mg of vitamin C per kg of diet, which helps strengthen their immune system and overall health.

Lipids are another essential component of the freshwater prawn's diet, requiring about 5% of lipids in their feed. A well-balanced diet that meets all their nutritional requirements is essential for freshwater prawns' overall health and growth.

Table 1.1: Nutritional requirement for M.rosenbergii diet

Protein	35-40%
Energy level	3.2Kcal/g feed
Protein-energy ratio	125-130 mg protein/kcal
Dietary fibers	30%
Vitamins	60-150mg vitamin C/kg diet
lipids	5%

Freshwater prawns can be reared in commercial farms using feeds that typically contain a protein content of 28-32%. While this protein content may seem limited, it is still adequate for the prawns' growth and development. Commercial feeds are often formulated to provide the necessary nutrients and minerals to support the prawns' health and growth, making them a viable option for farmers looking to rear freshwater prawns. With proper feeding and management practices, freshwater prawns can reach market size within a few months, making them an attractive option for aquaculture production.

3.2 Culture system:

The culture system plays a crucial role in the growth and development of Macrobrachium species. The type and design of the system, along with the specific parameters used, determine the overall yield of the species. By carefully selecting and optimizing these parameters, it is possible to enhance the growth pattern of Macrobrachium, leading to a higher yield and better quality of the species. For example, factors such as temperature, pH, water quality, feeding, and stocking density can all significantly impact the growth and survival of Macrobrachium in the culture system. Therefore, it is essential to consider and adjust these parameters according to the specific requirements of the species and the culture system to achieve optimal growth and yield.

3.3 Monosex culture:

3.3.1. All-male production system:

Prawns exhibit a hierarchical social structure characterized by three distinct morphotypes, each with unique characteristics. The morphotypes include the blue-clawed, orange-clawed, and petite males. The structure is such that the larger males tend to dominate over the smaller males, resulting in lower production prospects. This dominance is demonstrated through aggressive behavior, with larger males attacking and suppressing smaller males. This behavior often leads to a well-defined hierarchical relationship among males, with dominant males occupying the top position and smaller males occupying the lower positions.

Compared to females, males tend to develop a more structured hierarchical relationship. The females do not exhibit the same level of dominance, and their social interactions are relatively peaceful. The hierarchical structure among males, coupled with aggressive behavior, makes it necessary to keep the stocking density of prawns low to avoid overcrowding and reduce the incidence of aggressive behavior. This helps to maintain a healthy stock and ensure optimal production levels.

3.3.2. All-female production system:

According to a study conducted by Malecha et al. in 2010, it was found that all-female production significantly increased profit when compared to male and mixed-sex ratios. This could be attributed to female individuals exhibiting less hierarchical behavior and having comparatively lower levels of agonistic behavior. Additionally, all-female populations can be cultured in mass densities, making them a feasible option for aquaculture practices such as shrimp farming.

3.4 Green water culture:

Green water is a term used to describe a type of water system characterized by impounded water bodies containing high-density plankton populations. The main objective of this type of system is to increase natural feeding while reducing the need for supplemental feeding, as plankton provides 90% of the food required. This system is particularly effective in scampi larva culture, with the highest survival rate. However, this technique is still under development despite its effectiveness and has yet to be fully adopted for freshwater prawn culture.

The development of the green water system has not been without challenges. While studying larva culture in a green water static system in Malaysia, Anderson reported mass mortalities of larvae, indicating that the technique still needs to be refined to improve its effectiveness. Nevertheless, the potential benefits of the green water system, such as reduced feeding costs and improved survival rates, make it a promising technique for aquaculture.

In conclusion, the green water system is a unique and innovative approach to aquaculture that has shown promising results in scampi larva culture. While the technique is still developing, its potential benefits make it a technique worth exploring further in aquaculture.

3.5 Biofloc technology:

The study by Prajith K K et al. in 2010 delves into the use of biofloc in the larval rearing of Macrobrachium rosenbergii. Biofloc refers to using blue-green algae and beneficial bacteria to act as a feed supplement and detrivore, eliminating ammonia from the culture system. The study shows that the culture of macrobrachium prawn can be carried out until the larval rearing stages only on the biofloc. However, it must be noted that the steps involved in this process can be complex and may require further experimentation and refinement for future exploitation. Using biofloc has emerged as a promising technique in aquaculture and can potentially revolutionize how we approach sustainable and efficient aquaculture practices.

3.6 Composite culture

3.6.1 With tilapia:

The case study conducted by Alfredo Gracia et al. in Puerto Rico focuses on the composite culture of freshwater prawns and tilapia (Oreochromis niloticus) in inland waters. The study concludes that a ratio of seven parts freshwater prawn to one part tilapia is the most economically feasible and successful approach for farmers. Compared to monoculture, the yield for farmers is significantly higher in composite culture. It is worth noting that the minor species in the composite culture also play a crucial role in economic terms. Overall, the study highlights the importance of composite culture in achieving economic viability and sustainability in aquaculture.

3.6.2   With IMC:

A recent study by S. Jasmin et al. evaluated the economic feasibility of a composite culture of Indian Major Carps (IMC) with freshwater prawns in Bangladesh's water culture system. The researchers found that stocking M.rosenbergii at a density of 15000 per hectare acre is the most optimal approach. Furthermore, the study highlights that excluding bottom feeders like common carp can make the culture system more profitable. The findings of this study provide valuable insights for aquaculture practitioners and policymakers looking to enhance the economic viability of freshwater prawn and IMC farming in Bangladesh.

4. Resource and productivity:

In a research study by Karplus and colleagues, the agonistic behavior and growth suppression among three different morphotypes of scampi (Small male, orange-clawed male, and blue-clawed male) were investigated. The findings indicated that the larger blue-clawed males exhibit a dominant behavior and suppress the growth of other morphotypes, thereby establishing a clear hierarchy of morphotypes. BC is at the top, followed by OC and SM. All three morphotypes exist within the same social groups, and the suppression of smaller morphotypes by the larger ones is a common occurrence. The study sheds light on scampi's social dynamics and growth patterns, with implications for their overall yield.

5. Water quality parameters:

According to the research conducted by P.M. Sherry and colleagues, the appropriate water quality parameters for thriving scampi culture are essential. The following table lists the optimal levels of each parameter that should be maintained to ensure healthy growth and development of scampi.

Water temperature is one of the most critical factors when culturing scampi. The temperature should be maintained between 10-33°C, which is the most suitable for scampi culture.

Dissolved oxygen (DO) is another crucial parameter that should be closely monitored. The optimal DO range for monoculture of scampi is between 4.0 and 9.0 mg/L. A lower DO level can cause stress and even death, while a higher level can lead to reduced growth rate and reproduction.

pH is also an essential factor affecting scampi's survival and growth. The best pH range for scampi culture is 6.6-7.8. Any significant deviation from this range can cause damage to the scampi's respiratory system and affect its overall physiology.

Total alkalinity is another critical parameter that should be maintained within the 150-250 mg/L range. Alkalinity levels that are too low can cause a sudden drop in pH, while high alkalinity levels can cause the pH to increase, leading to sudden drops in DO levels.

Water hardness should also be monitored to ensure a thriving scampi culture. The optimal range of water hardness for scampi culture is between 110-160 mg/L. Hardness levels outside this range can affect the scampi's survival rate and reproductive success.

Phosphate is another critical parameter that can impact the growth and development of scampi. The best phosphate level for scampi culture is between 0.74 and 1.2 mg/L. High phosphate levels can cause algae blooms in the water, leading to a decrease in DO levels and affecting the growth of scampi.

Lastly, nitrate levels should be maintained within a 1.4-1.44 mg/L range, which is optimal for scampi culture. High levels of nitrate can cause stress and even mortality, while low levels can lead to reduced growth rate and reproductive success.

Table 5.1 Optimum water quality parameters for culturing freshwater prawn

Parameter	Optimum limit(For monoculture)
Water temperature (oC)	10–33
DO (mg/l)	4.0–9.0
pH	6.6–7.8
Total alkalinity (mg/l)	150–250
Hardness (mg/l)	110–160
Phosfate (mg/l)	0.74–1.2
Nitrate (mg/l)	1.4–1.44

6. Agonistic behavior and social growth suppression:

The study conducted by Karplus et al. provides a detailed analysis of agonistic behavior and growth suppression among three morphotypes of scampi: Small Male (SM), orange-clawed male (OC), and Blue-clawed Male (BC). The study found that the larger blue-clawed males tend to suppress the growth of the other two traits, making the hierarchy of morphotypes BC > OC > SM.

All three traits coexist in the same social groups, and the suppression of smaller ones is predominant. This suppression of more minor traits leads to an overall fluctuation in yield and lesser growth expectancy of scampi. The blue-clawed males take the food and reproductive privilege while suppressing the other two traits. Similarly, when orange-clawed and small-clawed males are in groups, the orange-clawed male tends to suppress the small-clawed males.

The study recommends "culling" by separating or killing the small-clawed prawns to ensure a uniform yield. This process is labor-intensive but necessary to ensure that the yield is consistent. The study also found that suppressing the more minor traits is due to the larger ones' ability to claim territory and resources, leading to a natural selection process favoring the more prominent traits.

In conclusion, the study provides valuable insights into the behavior and growth patterns of different morphotypes of scampi and recommends a culling process to ensure a consistent yield.

7. Mechanism of social growth control in adults and juveniles:

In a study by Ilan Karplus and colleagues, the growth suppression phenomenon in mature Macrobranchium rosenbergii males was explored and explained. The researchers found that a second pair of claws in blue-clawed males, known as "runts," was a crucial factor in controlling the growth of smaller-clawed males.

Further research has identified three primary mechanisms responsible for the growth control of freshwater prawns. The first mechanism is competition for food. Larger orange-clawed males tend to dominate over smaller blue-clawed and petite-clawed males, resulting in an uneven food distribution. This can lead to stunted growth in smaller males who cannot compete effectively for food resources.

The second mechanism is food intake suppression. Even if food is available, smaller males tend to have lesser food intake due to the size hierarchy. This means the larger males receive the lion's share of the available food resources, leaving less for the smaller males.

The third mechanism is decreased food conversion efficiency. Petite-clawed males usually have a lower food conversion ratio than other males, meaning they need to consume more food to achieve the same level of growth. This can lead to a situation where smaller males consume more food than larger males but still do not achieve the same level of growth, further exacerbating the growth control mechanism.

8. Standard stocking density:

The topic of stocking density is a significant point of debate among farmers. It is a delicate balance between under-stocking and over-stocking of cropland. Overcrowding can lead to the rapid spread of diseases and an increase in mortality rates, while under-stocking can result in under-utilization of cultivable land and lower crop yields. This segment will delve into the recommended stocking density for Macrobranchium rosenbergii de man, a freshwater prawn species commonly farmed in Asia.

A study conducted by Ranjeeth K et al. aimed to establish the standard stocking density for this species. The researchers experimented with different stocking densities and measured their effects on prawn growth and yield. The results showed that a density of 2.5 prawns per square meter, equivalent to 199.7 kg per hectare, yielded the maximum amount of crops. This density was more effective than both high and low stocking densities.

The study also noted that over-stocking led to an increase in the spread of diseases and a decrease in the overall yield. On the other hand, understocking resulted in a lower yield due to the under-utilization of available land. Therefore, farmers must maintain a proper stocking density to maximize their crop yields and maintain the health of their crops.

In conclusion, the recommended stocking density for Macrobranchium rosenbergii de man is 2.5 prawns per square meter. This density balances under-stocking and over-stocking, resulting in optimal yield and healthy crops.

9. Length-weight relation:

The length-weight relationship is usually used to relate the weight of a fish to its length; we can find one parameter if we know another. These two are usually related as,

$$W= aL^b$$

where W is the body weight of a prawn

L is the length of a prawn

a is intercept

 b is the growth coefficient

Rajeevan R et al. conducted a study on the growth patterns of Macrobranchium rosenbergii, in which they observed significant length and weight differentiation. The study found that the growth coefficient b varied from 2.5315, which allowed for the determination of the value of a. By using the values of a and b, it became possible to find the relative weight of the species in terms of its growth. This study sheds light on the growth patterns of Macrobranchium rosenbergii and provides valuable insights into the factors that affect its development.

10. Practical alternate feed:

Feeding plays a crucial role in the successful rearing of prawns. Adequately balanced feed is essential for ensuring their survival and maintaining good health. Practical alternate feeds have gained popularity as a substitute for formulated feeds. These practical feeds often use plankton and other natural sources of nutrition that have similar properties to formulated feeds. By utilizing practical alternate feeds, prawn farmers can reduce their dependence on costly formulated feeds while still providing the necessary nutrients for the prawns to thrive. However, it is essential to ensure that the practical feeds used are correctly balanced and meet the nutritional needs of the prawns at different stages of their growth.

10.1 Enriched copepods:

N.W. Rasdi's research has demonstrated the efficacy of using enriched copepods as an alternative source of nutrition for scampi. In this study, the copepod species Oithona sp. was enriched with mixed algae (Tetraselmis et al.) and fed to the prawns twice a day at a rate of 6-7% of their body weight. The study's results indicate that the prawns' survival rate was exceptionally high, at 95.6 -+ 1.04%, with a growth rate of 4.96±0.02%. These findings suggest that using enriched copepods could be a promising method for enhancing the growth and survival of scampi in aquaculture settings.

10.2 Microalgae:

Malwine Lober, a researcher, conducted a study on the feasibility of using microalgae as an alternate source of feed for scampi. In her study, she used a specific microalgae species called Nanochloropsis sp. and found that the survival rate of scampi fed with this microalgae was 70.8%. Additionally, the study reported that the growth rate of the scampi was 3.03+-0.44^a, where "a" represents the intercept. This information shows that using microalgae as an alternate source of feed for scampi could be a viable option for aquaculture.

10.3 Biofloc technology:

Flock is a term used to describe the in-situ production of microorganisms that convert the ammonia accumulated from fish excretion into food for other organisms. This process is known as bio floc technology, which involves manipulating the microbial community in aquaculture systems to reduce water exchange and improve water quality. J. Alberto P´erez-Fuente extensively studied applying bio floc technology in freshwater prawn culture. Using a bio floc system in cultivation has shown remarkable results, with an 85% survival rate achieved at 37 PL /m2 stocking density. Biofloc technology has been proven to be an effective and sustainable aquaculture method that can improve the overall health and yield of aquatic organisms while reducing the environmental impact of aquaculture.

Comparison chart:

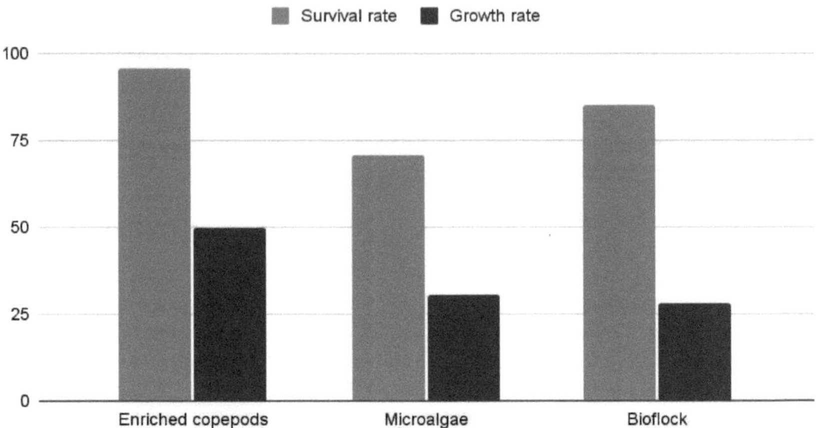

Figure 2 Comparison of growth performance by giant freshwater prawn in various culture systems

11. Influence of temperature on growth and reproduction:

Freshwater prawns are highly sensitive to temperature changes, and the water temperature can significantly impact their growth and development. According to research conducted by S.M. Manush and A.K. Pal in 2006, the ideal temperature range for their development is between 25 to 36 degrees Celsius, which is considered ambient. However, temperatures exceeding this range can adversely affect their growth and development.

The study also found that the time it takes for the larva to mature increases as the temperature goes beyond the ideal range, and the reverse is also true when the temperature drops below the optimum range. Temperature is crucial in freshwater prawns' developmental rates and morphometrics.

Moreover, the hatching of eggs and their maturation are also temperature-dependent. The ideal temperature for their hatching is between 27 to 30 degrees Celsius. If the temperature drops below this range, the eggs may not hatch, and if the temperature goes beyond this range, it can lead to abnormal development in the larvae.

Additionally, mating and spawning are also restricted to specific temperature ranges. The ideal temperature range for mating and spawning is between 25 to 28 degrees Celsius. If the temperature exceeds this range, it can reduce mating behavior and spawning activity.

In conclusion, maintaining the ideal temperature range is essential for freshwater prawns' growth, development, and reproduction.

12. Neo-female technology:

The seed production process of scampi is delicate and requires much attention to detail. One of the challenges faced during the process is the possibility of sex reversal of male to female due to the removal of androgenic glands. However, despite this challenge, the all-male population is still economically superior to mixed and all-female populations.

A study conducted by Wikrom Rungsin further confirmed this fact. The study found that springs produced by no females are predominantly male, which is significant from an economic standpoint. Additionally, the fecundity of no females was found to be as high as 4000 to 25000 eggs, which is quite impressive.

Overall, the process highlights the importance of maintaining the all-male population during the seed production of scampi. This will not only help to ensure economic viability but also contribute to the overall success of the process.

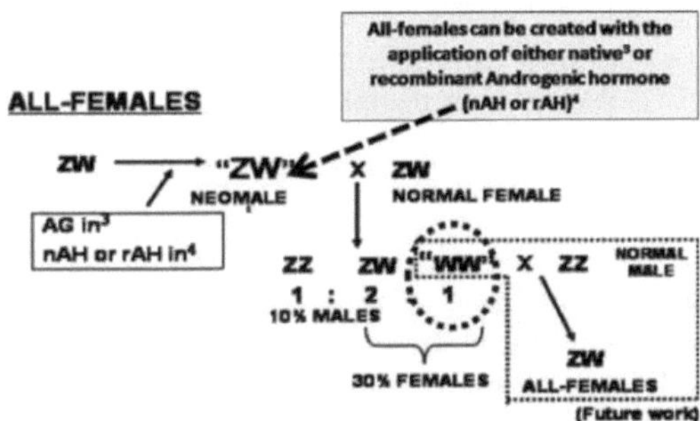

Figure 3 Neo-female technology for producing all-male freshwater prawns

13. Utilizing processing discards:

Processing discards refer to the parts of shellfish separated from their edible flesh. These include the head, carapace, shells, and appendages such as leggings and antennae. These discards are usually considered waste products and are thrown away, but there is a better way to use them.

One of the ways to utilize these discards is by processing them into various useful products. For instance, crustaceans' shells contain chitin, a natural polymer. Chitin has many applications, such as in textile printing, ion exchange chromatography, and treating industrial effluents.

The process of obtaining chitin involves removing the proteins from the shellfish shells through a process known as demineralization. The remaining chitin is then further processed into chitosan and has several uses. For example, chitosan can be used as a food additive to enhance the shelf life of food products. It can also be used as a co-material for packaging materials, which makes it an environmentally friendly alternative to plastic.

Another essential polysaccharide that can be prepared from shellfish discards is glucosamine. Glucosamine is a natural compound used as a food supplement and in treating osteoarthritis. It is also known to have anti-inflammatory properties.

Using shellfish discards to produce valuable products is an innovative way to contribute to the farming economy. Proper processing techniques can transform these discards into valuable materials with various industrial, environmental, and health applications.

14. Larviculture advancements:

In hatcheries, the survival of scampi larvae is of utmost importance, and infections such as vibriosis can cause significant mortality rates. To address this issue, a study was conducted to investigate using Poly β-hydroxybutyrate (PHB) as a feed for Artemia nauplii Instar II stage. The study found that PHB effectively inhibited the growth of various bacteria, including vibrio, which is a significant cause of larval mortality. Moreover, the larvae easily consumed PHB, resulting in reduced mortality rates. The incorporation of PHB into the feed of Artemia nauplii Instar II stage is a promising solution to the problem of larval mortality in scampi hatcheries.

Conclusion:

According to M. Mathiesen, the Assistant Director-General for Fisheries and Aquaculture at the United Nations Food and Agriculture Organisation (FAO), approximately 70% of our planet is covered by oceans. However, this vast expanse provides only 2% of our food. He believes there is immense potential for aquaculture resources to feed and sustain the livelihoods of our growing population, but this must be done sustainably. Mathiesen stresses the importance of ensuring that our exploitation of oceanic resources does not lead to their depletion or destruction. In his view, there is significant potential for extending scampi culture, which could help meet our food demands while promoting sustainable fishing practices.

Reference:

1. FAO. 2020. The State of World Fisheries and Aquaculture 2020. Sustainability in action. Rome. https://doi.org/10.4060/ca9229en

2. New, M. B., S. Singholka and M.N. Kutty. 2000. Prawn captures fisheries and enhances fisheries. Pages 411-428 In M.B. New and W. Valenti, editors, Freshwater Prawn Culture: The Farming of Macrobrachium rosenbergii. Blackwell Science, Oxford, UK

3. Saurabh, S. and Sahoo, P.K., 2008. Major diseases and the defense mechanism in giant freshwater prawn, Macrobrachium rosenbergii (de Man). *Proceedings of the National Academy of Sciences, India, Section B, 78*, pp.103-121.

4. Lalrinsanga, P.L., Pillai, B.R., Patra, G., Mohanty, S., Naik, N.K. and Sovan, S.A.H.U., 2012. Length-weight relationship and condition factor of giant freshwater prawn Macrobrachium rosenbergii (De Man, 1879) based on developmental stages, culture stages, and sex. *Turkish Journal of Fisheries and Aquatic Sciences, 12*(4).

5. Rungsin, W., Paankhao, N. and Na-Nakorn, U., 2006. Production of all-male stock by neofemale technology of the Thai strain of freshwater prawn, Macrobrachium rosenbergii. *Aquaculture, 259*(1-4), pp.88-94.

6. Aflalo, E.D., Dandu, R.V., Verghese, J.T., Rao, N., Samraj, T.Y., Ovadia, O. and Sagi, A., 2014. Neo-females production and all-male progeny of a cross between two Indian strains of prawn (Macrobrachium rosenbergii): Population structure and growth performance under different harvest strategies. *Aquaculture, 428*, pp.7-15.

7. Manush, S.M., Pal, A.K., Das, T. and Mukherjee, S.C., 2006. The influence of 25 to 36 C temperatures on developmental rates, morphometrics, and survival of freshwater prawn (Macrobrachium rosenbergii) embryos. *Aquaculture, 256*(1-4), pp.529-536.

8. Malecha, S., 2012. The case for all-female freshwater prawn, M acrobrachium rosenbergii (D e M an), culture. *Aquaculture Research, 43*(7), pp.1038-1048.

9. Mohanakumaran Nair, C., Salin, K.R., Raju, M.S. and Sebastian, M., 2006. Economic analysis of mono sex culture of giant freshwater prawn (Macrobrachium et al.): a case study. *Aquaculture Research, 37*(9), pp.949-954.

10. Sagi, A. and Aflalo, E.D., 2005. The androgenic gland and monosex culture of freshwater prawn Macrobrachium rosenbergii (De Man): a biotechnological perspective. *Aquaculture Research, 36*(3), pp.231-237.

11. Aflalo, E.D., Hoang, T.T.T., Nguyen, V.H., Lam, Q., Nguyen, D.M., Trinh, Q.S., Raviv, S. and Sagi, A., 2006. A novel two-step procedure for mass production of all-male populations of the giant freshwater prawn Macrobrachium rosenbergii. *Aquaculture, 256*(1-4), pp.468-478.

12. Thanh, N.M., Ponzoni, R.W., Nguyen, N.H., Vu, N.T., Barnes, A. and Mather, P.B., 2009. Evaluation of growth performance in a diallel cross of three strains of giant freshwater prawn (Macrobrachium rosenbergii) in Vietnam. *Aquaculture, 287*(1-2), pp.75-83.

13. Nhan, D.T., Wille, M., De Schryver, P., Defoirdt, T., Bossier, P. and Sorgeloos, P., 2010. The effect of poly β-hydroxybutyrate on larviculture of the giant freshwater prawn Macrobrachium rosenbergii. *Aquaculture, 302*(1-2), pp.76-81.

14. Karplus, I., 2005. Social control of growth in Macrobrachium rosenbergii (De Man): a review and prospects for future research. *Aquaculture Research, 36*(3), pp.238-254.

15. New, M.B. and Nair, C.M., 2012. The global scale of freshwater prawn farming. *Aquaculture Research, 43*(7), pp.960-969.

16. Gupta, A., 2005. *Formulation and evaluation of some alternative practical feeds for giant freshwater prawn, Macrobrachium rosenbergii (De Man)* (Doctoral dissertation, Punjab Agricultural University; Ludhiana).

17. Sharma, A., 2002. *OPTIMIZATION OF PLANT TO ANIMAL PROTEIN RATIO IN THE POST LARVAL DIET OF MACRO BRACHIUM ROZENBERGII* (Doctoral dissertation).

18. Ranjeet, K. and KURUP, B.M., 2011. We are standardizing stocking density for freshwater prawn Macrobrachium rosenbergii (De Man, 1879) farming in coconut garden channels. *Asian Fisheries Science, 24*(4), pp.354-366.

19. Uddin, M.S., Azim, M.E., Wahab, M.A. and Verdegem, M.C.J., 2009. Effects of substrate addition and supplemental feeding on plankton composition and production in tilapia (Oreochromis niloticus) and freshwater prawn (Macrobrachium rosenbergii) polyculture. *Aquaculture, 297*(1-4), pp. 99-105.

20. Farming, M., Towards Monosex Culture of Prawns (Scampi).

21. Singh, S. and Chauhan, R.S., 2014. Study on Growth and Survival of Giant Freshwater Prawn, Macrobrachium rosenbergii, in Tarai Agroclimatic Regime of Uttarakhand. *Journal of Ecophysiology and Occupational Health, 14*(3-4), pp.165-170.

22. Adhikari, S., Sahu, B.C. and Dey, L., 2012. Nutrient budget and effluent characteristics in polyculture of scampi (Macrobrachium rosenbergii) and Indian central carp ponds using organic inputs. *Water Science and Technology, 66*(7), pp.1540-1548.

23. Ranjeet, K. and Kurup, B.M., 2011. Density depends on variations in the production and population structure of Macrobrachium rosenbergii reared in the wetland polders of south India. *Brazilian Journal of Aquatic Science and Technology, 15*(2), pp.55-62.

24. Jena, A.K., Biswas, P. and Saha, H., 2017. Advanced farming systems in aquaculture: strategies to enhance the production. *Innovative Farming, 1*(1), pp.84-89.

25. Wasave, S.S., Singh, H. and Wasave, S.M., 2019. Cashew nut Anacardium occidentale L. byproducts as an alternative protein source for post-larvae of Macrobrachium rosenbergii (de Man, 1879).

26. Prajith, K.K. and Madhusoodana, K.B., 2011. *Application of Biofloc Technology (BFT) in the nursery rearing and farming of giant freshwater prawn, Macrobrachium rosenbergii (deMan)* (Doctoral dissertation, Cochin University of Science and Technology).

27. Ghosh, R., Banerjee, K., Homechaudhuri, S. and Mitra, A., 2010. Effect of formulated feeds on freshwater prawn growth and water quality (Macrobrachium rosenbergii) farming: A case study from Gangetic Delta—Livestock *Research for Rural Development, 22*(12).

28. Jeyanthi, P. and Gopal, N., 2012. Growth and instability in Indian frozen scampi export.

29. Zynudheen, A.A., 2020. Utilization of shellfish processing discards. ICAR-Central Institute of Fisheries Technology.

30. Suday, P. and Hardial, S., 2006. Maturity indicators for freshwater prawn crops. *Journal of Applied Zoological Researches, 17*(1), pp.103-105.

31. Balamurugan, P., Mariappan, P. and Balasundaram, C., 2004. Impacts of mono-sex Macrobrachium culture on the future of seed availability in India. *Aquaculture Asia, 9*, pp.15-16.

32. Jaiswal, K., Vadher, K.H., Pal, M., Manusha, K. and Sharman, V., 2017. NUTRITIONAL ASPECT FOR FRESHWATER PRAWN (Macrobrachium rosenbergii) FARMING. *Editorial Board, 6*(8), p.217.

33. Jena, A.K., Biswas, P. and Saha, H., 2017. Advanced farming systems in aquaculture: strategies to enhance the production. *Innovative Farming, 1*(1), pp.84-89.

34. Arya, P., Chauhan, R.S., Khati, A. and Nazir, I., Organic aquaculture: A boon to the fisheries industry.

35. Maddox, M.B. and Manzi, J.J., 1976, March. The Effects Of Algal Supplements On Static System Culture Of Macrobrachium rosenbergii (De Man) Larvae 1, 2. In *Proceedings of the annual meeting-World Mariculture Society* (Vol. 7, No. 1-4, pp. 677-698). Oxford, UK: Blackwell Publishing Ltd.

36. Lober, M. and Zeng, C., 2009. Effect of microalgae concentration on larval survival, development and growth of an Australian strain of giant freshwater prawn Macrobrachium rosenbergii. *Aquaculture, 289*(1-2), pp.95-100.

37. Arya, P., Chauhan, R.S., Khati, A. and Nazir, I., Organic aquaculture: A boon to the fisheries industry.

38. Maddox, M.B. and Manzi, J.J., 1976, March. The Effects Of Algal Supplements On Static System Culture Of Macrobrachium rosenbergii (De Man) Larvae 1, 2. In *Proceedings of the annual meeting-World Mariculture Society* (Vol. 7, No. 1-4, pp. 677-698). Oxford, UK: Blackwell Publishing Ltd.

39. Lober, M. and Zeng, C., 2009. Effect of microalgae concentration on larval survival, development and growth of an Australian strain of giant freshwater prawn Macrobrachium rosenbergii. *Aquaculture, 289*(1-2), pp.95-100.

40. Tan, K., Jiang, H., Jiang, D. and Wang, W., 2020. Sex reversal and the androgenic gland (AG) in Macrobrachium rosenbergii: A review. *Aquaculture and Fisheries, 5*(6), pp.283-288.

41. Rasdi, N.W., Ikhwannuddin, M., Syafika, C.A., Azani, N. and Ramli, A., 2021. Effects of using enriched copepod with microalgae on growth, survival, and proximate composition of giant freshwater prawn (Macrobrachium rosenbergii). *Iranian Journal of Fisheries Sciences, 20*(4), pp.986-1003.

42. Kruangkum, Thanapong, Jirawat Saetan, Charoonroj Chotwiwatthanakun, Rapeepun Vanichviriyakit, Sirorat Thongrod, Pakpoom Thintharua, Tatpong Tulyananda, and Prasert Sobhon. "Co-culture of males with late premolt to early postmolt female giant freshwater prawns, Macrobrachium rosenbergii resulted in greater abundances of insulin-like

androgenic gland hormone and gonad maturation in male prawns as a result of olfactory receptors." *Animal Reproduction Science* 210 (2019): 106198.

43. Azad, M.A.K., Islam, S.S., Amin, M.N., Ghosh, A.K., Hasan, K.R., Bir, J., Banu, G.R. and Huq, K.A., 2021. Production and economics of probiotics treated Macrobrachium rosenbergii at different stocking densities—Animal *Feed Science and Technology, 282*, p.115125.

44. Sumon, M.S., Ahmmed, F., Khushi, S.S., Ahmmed, M.K., Rouf, M.A., Chisty, M.A.H. and Sarower, M.G., 2018. Growth performance, digestive enzyme activity, and immune response of Macrobrachium rosenbergii fed with probiotic Clostridium butyricum incorporated diets. *Journal of King Saud University-Science, 30*(1), pp.21-28.

45. Jasmine, S., Molina, M., Hossain, M.Y., Jewel, M.A.S., Ahamed, F. and Fulanda, B., 2011. Potential and economic viability of freshwater prawn Macrobrachium rosenbergii (de Man, 1879) polyculture with significant Indian carp in Northwestern Bangladesh. *Our Nature, 9*(1), pp.61-72.

46. Karplus, I., Barki, A. and Goren, M., 1991. The agonistic behavior of the three male morphotypes of the freshwater prawn Macrobrachium rosenbergii (Crustacea, Palaemonidae). Behavior, 116(3-4), pp.252-276.

47. Segal, E. and Roe, A., 1975. In close confinement, the growth and behavior of post-juvenile Macrobrachium rosenbergii (de Man). Proc. World Mariculture. Soc.. 6: 67-88.

48. Cobb, J.S.. Tamm, G.R. and Wan&D.: 1982. Behavioral mecha&ns influ&ng molt frequency in the American lobster Homarus americanus Milne Edwards. J. Exp. Mar. Biol. Ecol., 62: 185-200.

49. Karplus, I., Hulata, G. and Zafrir, S., 1992. Social control of growth in Macrobrachium rosenbergii. IV. The mechanism of growth suppression in runts. *Aquaculture, 106*(3-4), pp.275-283.

50. Ranjet, K. and Kurup, B.M., 2013. Economic analysis of polder-based freshwater prawn farming systems in Kuttanad, India. *Int. J. Fish. Aquacult., 5*, pp.110-121.

51. Rajeevan, R., Edirisinghe, U. and Athauda, A.R.S.B., 2018. Length-weight relationships and condition factor of giant freshwater prawn, Macrobrachium rosenbergii (De Man, 1879) in five perennial reservoirs in Northern Province, Sri Lanka. *International Journal of Fisheries and Aquatic Studies, 6*(5), pp.283-287.

52. Rasdi, N.W., Ikhwannuddin, M., Syafika, C.A., Azani, N. and Ramli, A., 2021. Effects of using enriched copepod with microalgae on growth, survival, and proximate composition of giant freshwater prawn (Macrobrachium rosenbergii). *Iranian Journal of Fisheries Sciences, 20*(4), pp.986-1003.

YOUR KNOWLEDGE HAS VALUE

- We will publish your bachelor's and
  master's thesis, essays and papers

- Your own eBook and book -
  sold worldwide in all relevant shops

- Earn money with each sale

Upload your text at www.GRIN.com
and publish for free